Pezhman Ghadimi
Mohammadreza Khoei

Lean Supply Chain Management Information Modeling

Misam Kashefi
Pezhman Ghadimi
Mohammadreza Khoei

Lean Supply Chain Management Information Modeling

Value Stream Mapping(VSM)

LAP LAMBERT Academic Publishing

Impressum/Imprint (nur für Deutschland/only for Germany)
Bibliografische Information der Deutschen Nationalbibliothek: Die Deutsche Nationalbibliothek verzeichnet diese Publikation in der Deutschen Nationalbibliografie; detaillierte bibliografische Daten sind im Internet über http://dnb.d-nb.de abrufbar.
Alle in diesem Buch genannten Marken und Produktnamen unterliegen warenzeichen-, marken- oder patentrechtlichem Schutz bzw. sind Warenzeichen oder eingetragene Warenzeichen der jeweiligen Inhaber. Die Wiedergabe von Marken, Produktnamen, Gebrauchsnamen, Handelsnamen, Warenbezeichnungen u.s.w. in diesem Werk berechtigt auch ohne besondere Kennzeichnung nicht zu der Annahme, dass solche Namen im Sinne der Warenzeichen- und Markenschutzgesetzgebung als frei zu betrachten wären und daher von jedermann benutzt werden dürften.

Coverbild: www.ingimage.com

Verlag: LAP LAMBERT Academic Publishing GmbH & Co. KG
Dudweiler Landstr. 99, 66123 Saarbrücken, Deutschland
Telefon +49 681 3720-310, Telefax +49 681 3720-3109
Email: info@lap-publishing.com

Approved by: Johor Bahru, University Technology Malaysia,2011

Herstellung in Deutschland:
Schaltungsdienst Lange o.H.G., Berlin
Books on Demand GmbH, Norderstedt
Reha GmbH, Saarbrücken
Amazon Distribution GmbH, Leipzig
ISBN: 978-3-8454-2966-3

Imprint (only for USA, GB)
Bibliographic information published by the Deutsche Nationalbibliothek: The Deutsche Nationalbibliothek lists this publication in the Deutsche Nationalbibliografie; detailed bibliographic data are available in the Internet at http://dnb.d-nb.de.
Any brand names and product names mentioned in this book are subject to trademark, brand or patent protection and are trademarks or registered trademarks of their respective holders. The use of brand names, product names, common names, trade names, product descriptions etc. even without a particular marking in this works is in no way to be construed to mean that such names may be regarded as unrestricted in respect of trademark and brand protection legislation and could thus be used by anyone.

Cover image: www.ingimage.com

Publisher: LAP LAMBERT Academic Publishing GmbH & Co. KG
Dudweiler Landstr. 99, 66123 Saarbrücken, Germany
Phone +49 681 3720-310, Fax +49 681 3720-3109
Email: info@lap-publishing.com

Printed in the U.S.A.
Printed in the U.K. by (see last page)
ISBN: 978-3-8454-2966-3

ACKNOWLEDGEMENT

I express my colleagues and friends who helped me to prepare and improve this book. The contributions of Mr. Pezhman Ghadimi, Mr. Mohammadreza Khoei and Elena Alexei were particularly valuable. My colleagues have provided intellectually stimulating environments in which to work. The contributions of the professional practitioners with whom I have worked have been invaluable. I am indebted to Lap-Publishing and the editors for permission to use copyrighted material. I'm grateful to Muda and MC Pack for let me to access to their information database and also UTM University for financial supporting of this project.

ABSTRACT

Lean manufacturing uses less of everything compared to mass production i.e. reduce in human effort, manufacturing space, investments on tools and equipment, processing time that its popularity has shown itself as an important production method in industries. There have been numerous reports of the positive outcomes enjoyed by manufacturers result of implementing lean manufacturing.

This report provides a value stream mapping (VSM) of a printer box production where causes that lead to waste are identified and minimized using the lean manufacturing technique. The study defined, measured and analyzed problems in the production. Improvements alternatives are proposed to improve the production line using eVSM software where the current VSM and that of the new production line are established. Differences between them are identified where the benefits of the later are highlighted and suggested to the company

Results show that current productions have: 4 days for pallet, 3.9 days for printing parts, 5.88 days for die-cut/ stitching/gluing work in progress, 2.85 days finishing goods and 17 days production lead time. Improvement with leveling, there are noticeable reduction that shows : 1.5 days pallet, 1 day for printing parts, 0 day (negligible) for die-cut/ stitching/gluing work in progress, 2 days for finishing goods and 5 days for production lead time. Future works include mapping other production lines, develop a larger model of company, application to supply chain and quality management, use simulation software such as witness and simplify information flow using online work order for customer demand hence avoiding constraint.

TABLE OF CONTENTS

LIST OF TABLES

LIST OF FIGURES

CHAPTER 1

INTRODUCTION

1.1   Introduction

Manufacturers in the paper milling and packaging has always faced heightened challenges such as rising customers' expectation, fluctuating demand, and competition in markets. There is no doubt that these manufacturers have always embraced changes and improvements in their key activities or processes to cope with the challenges. One way to stay competitive in this globalized market is to become more efficient. By adopting lean manufacturing that has received a lot of attention among industrial sectors.

There have been numerous reports on the positive outcomes of implementing lean manufacturing among the paper milling packaging industries. Lean manufacturing uses less of everything compared to mass production,i.e.- half the human effort in the factory, half the manufacturing space, half the investment in tools, and half the engineering hours to develop a new product (Womack et al., 1990). It has now become a production method for many manufacturers to pursue.

Little studies regarding lean manufacturing have been done in Malaysia. A survey needs to be carried out in order to gauge how organizations in this country practice it. This research was initiated with a focus to examine the adoption of lean manufacturing in the paper milling and packaging industry. Various issues such as its understanding among the respondent companies, its benefits and obstacles, the tools and techniques used etc, were investigated. In addition, the degree of implementation of key practice areas of lean manufacturing was assessed.

This research begins with a general overview of lean manufacturing and the key areas that characterize its adoption. This is followed by an outline of the methodology employed for conducting the survey. Findings of the survey together with the results of some statistical analyses that were applied are presented in the next section. Finally the research ends with conclusions.

1.2   Background of the Project

Management information systems are progressively important to the prosperity of many companies (Chopra and Meindl 2001; Kumar 2001). Few researches have investigated the profits and abilities of various kinds of management information systems such as Electronic Data Interchange (EDI) (Lee et al. 1999; Mukhopadhyay et al. 1995), electronic market (Dagenais and Gautschi 2002; Kaplan and Sawhney 2000), or extended enterprise resource scheduling (Green 2001) systems.

Nevertheless, there are few experimentally extracted models appropriate for examining the scope of management information systems options. Similarly, companies face complicated and dangerous determination analyzing and choosing a suitable management information systems solution or guaranteeing that their executed systems are organized with their competitive policies (Reddy and Reddy 2001). While the improved policies suitable model is operationalized particularly for management information systems, the fundamental hypothesis and methodology could be accepted for analyzing other kinds of information systems.

1.3   Problem Statement

Information system for lean production issues in supply chain plays increasingly and vital role in the ability of factories to decrease costs and increase the performance of their supply chain. The conventional management methods, like the sequential method of project realization, traditional approach to quality and segmented control could be detrimental to construction flows. A number of solution or visions have been offered to relieve the problem in supply chain. One of them is lean production.

Lean production is a philosophy developed in manufacturing industry. It provides a new approach for solving the problem and profitably delivers what the customer needs. The concept of lean manufacturing (LM) for services has matched with process flow and waste issues. More attention, mainly in developing countries has paid to level of satisfaction in service sectors. So it seems necessary to find an optimized solution or system to achieve the most possible value-added tasks. In addition, an information system needs to be developed to facilitate the information flow of system and pave the way of making suitable, on-time and effective decisions through the whole supply chain.

The current study focused on the company A which adopted the traditional concept of push production system in their production line. This in turns created high level of work in progress, long lead time and low value added ratio. This project was carried out to address these problems.

1.4   Objective of Study

The objectives of the study are:

- To value stream map the current situation of the Printer Box Model Brother Production line.
- To identify waste using lean manufacturing technique and determine causes to the waste through.
- To reduce manufacturing lead time and develop a future state VSM based on the best improvement
- To develop IDEF1X and flow of information between supply chain members

1.5   Scope of Study

The study focused on the paper packaging of Brother Printer box Model. The study would define, measure, and analyze the problems and identify areas of improvement by analyzing the value stream maps. Improvement alternatives would be proposed by trying to improve the operations in the production line. To approach this aim we used Microsoft Access was used to model the data information relationship between supply chain members. The eVSM software to show the value stream mapping of the current and improved production line. The best improvement alternative would be suggested to Company A but implementation was not within the scope of the project due to time and financial constraints.

1.6   Significance of Study

Today high competitive market needs fast, effective, high responsiveness, online interactive, 24 hours 7 days availability, easy to follow up orders processing, etc. For researchers, and the model provides an interdisciplinary approach to understand the range of lean supply chain capabilities. With further study, the model and concepts could also be adapted for other strategic information system applications. Practitioners can gain a better understanding of the capabilities of their implemented lean supply chain and the expected capabilities that future lean supply chain may provide.

1.7   Structure of Thesis

This report consists of six chapters, as summarized below:

- **Chapter 1 Introduction**

Chapter 1 is the introduction of the study. This chapter explains about the research statement, problem statement, objective of study, scope of study and matters that have relate to the introduction of project.

- **Chapter 2 Literature Review**

Chapter 2 is the literature review of the project and contains on several topic related to this study, describe definition, principle and approach used in conducting this project. Topics reviewed include information system, information modeling, lean concept, seven wastes, value stream mapping, value added and non value added, concept supply chain, introduction to eVSM, Microsoft access approach, different methodologies used and comparison between them and finally the conclusion.

- **Chapter 3 Research Methodology**

Chapter 3 discusses on the methodology of how the gathered data regarding the case product would be analyzed by using eVSM approach with the Microsoft Access software.

- **Chapter 4 Case Study and Data Collection**

This chapter is about the collected information related to the product (Packing Printer Box) to be assessed.

- **Chapter 5 Results and Discussion**

Chapter 5 displays the result and data analysis that assess by the eVSM software. Also, there will be some discussions for each result.

- **Chapter 6 Conclusion**

Chapter 6 consists of a summary of whole study. Then, Findings of the research are presented in brief. Finally, some future research is suggested.

1.8   Conclusion

This chapter highlighted given a general introduction about the entire study. At the beginning of this chapter, the introduction of lean manufacturing and supply chain management were briefly discussed. It was followed by the research statement and the problems that this area is faced with. The objectives and scopes of the project were stated to address the boundaries of the study. The significance of the study was discussed. Lastly, the arrangement of the entire report was explained.

CHAPTER 2

LITERATURE REVIEW

2.1   Introduction

Womack et al. (1990) first introduced the term "lean production" in "The Machine that Changed the World". Where lean production was referred to the manufacturing system established by the Toyota Production System. Lean is a continual improvement process focused on eliminating waste and delivering quality products and services at the lowest cost to consumer. Table 2.1 shows the seven types of wastes in the lean context (Liker and Meier, 2006):

Table 2.1 : Type of waste

No.	Waste type	Description
1	Overproduction	Producing items earlier or in greater quantities than needed by the customer. Generates other wastes, such as overstaffing, storage, and transportation costs because of excess inventory. Inventory can be physical inventory or a queue of information.
2	Waiting	Workers merely serving as watch persons for an automated machine, or having to stand around waiting for the next processing step, tool, supply, part, etc., or just plain having no work because of no stock, lot processing delays, equipment downtime, and capacity bottlenecks.

No.	Waste type	Description
3	Transportation or conveyance	Moving work in progress (WIP) from place to place in a process, even if it is only a short distance. Or having to move materials, parts, or finished goods into or out of storage or between processes.
4	Over-processing or incorrect processing	Taking unneeded steps to process the parts. Inefficiently processing due to poor tool and product design, causing unnecessary motion and producing defects. Waste is generated when providing higher quality products than is necessary.
5	Excess inventory	Excess raw material, WIP, or finished goods causing longer lead times, obsolescence, damaged goods, transportation and storage costs, and delay. Also, extra inventory hides problems such as production imbalances, late deliveries from suppliers, defects, equipment downtime, and long setup times.
6	Unnecessary movement	Any motion employees have to perform during the course of their work other than adding value to the part.
7	Defects	Production of defective parts or correction. Repairing of rework, scrap, replacement production, and inspection means wasteful handling, time, and effort.

The basic idea of lean was to identify and eliminate wastes from every aspect of the business (Levinson and Rerick, 2002). The Japanese word for waste was "muda" and was classified into Type-1 muda (Non-value-added but necessary) and Type-2 muda (Non-value-added and not necessary) (Sayer and Williams, 2007). Hiroyuki Hirano defined waste as "everything that is not absolutely essential"(Santos, Wysk and Torres, 2006). In lean manufacturing, waste is the synonym of non-value added.

In lean manufacturing, when talks about Value-Added and Non-Value Added, they are all defined from the point of view of the customer. In order for an activity to be considered value-added, all actions, activities, processes, persons, organizations, systems, pieces of equipment, and any other resources committed to the process must meet these three criteria (Sayer and Williams, 2007):

✓ The customer must be willing to pay for the activity.

✓ The activity must transform the product or service in some way.

✓ The activity must be done correctly the first time.

Ross & Associates Environmental Consulting Ltd. (2003) mentioned in their report, that lean production typically represents a paradigm move from conventional "batch and queue" functionally aligned mass production to "one-piece flow" product- lined pull production. This shift requires the implementation of just-in time production principle and employee-involved, system-wide, continual improvement. Several potential outcomes of implementing lean are identified as follow:

1. **Reduced inventory levels** (raw material, work-in-progress, finished product) along with associated carrying costs and loss due to damage, spoilage, off-specification, etc;

2. **Decreased material usage** (product inputs, including energy, water, metals, chemicals, etc.) by reducing material requirements and creating less material waste during manufacturing;

3. **Optimized equipment** (capital equipment utilized for direct production and support purposes) using lower capital and resource intensive machines to drive down costs;

4. **Reduced need for factory facilities** (physical infrastructure primarily in the form of buildings and associated material demands) by driving down the space required for product production;

5. **Increased production velocity** (the time required to process a product from initial raw material to delivery to a consumer) by eliminating process steps, movement, wait times, and downtime;

6. Enhanced production flexibility (the ability to alter or reconfigure products and processes rapidly to adjust to customer needs and changing market circumstances) enabling the implementation of a pull production, just-in-time oriented system which lowers inventory and capital requirements; and

7. Reduced complexity (complicated products and processes that increase opportunities for variation and error) by reducing the number of parts and material types in products, and by eliminating unnecessary process steps and equipment with unneeded features.

2.2   Value Stream Mapping

A value stream map (VSM) is a static and graphical representation of how all the steps in any process line up to produce a product or service, as well as the flow of information between the processes (Sayer and Williams, 2007). Gidley (2004) offers a reference guideline on how to map current and future state VSM. Gidley thinks that current state VSM should be developed using the following structure:

1. Draw Customer, Supplier and Production Control icons
2. Enter customer requirements per month and per day
3. Calculate daily production and container requirements
4. Draw outbound shipping icon and truck with delivery frequency
5. Draw inbound shipping icon, truck and frequency
6. Add process boxes in sequence, left to right
7. Add data boxes below process boxes
8. Add communication arrows and note methods and frequencies
9. Obtain process attributes and add to data boxes. Observe all times directly.
10. Add operator symbols and numbers
11. Add inventory locations and levels in days of demand and graph at bottom

12. Add push, pull and FIFO icons

13. Add other information that may prove useful

14. Add working hours

15. Cycle and Lead times

16. Calculate Total Cycle Time and Lead Time

Develop a future state VSM needs more art, engineering, and strategy than to develop a current state VSM (Gidley, 2004). Gidley also suggested a structure to construct future state VSM:

1. Identify bottleneck processes, which is the operation with the longest cycle time

2. Identify lot sizing or setup reduction opportunities

3. Identify potential work cells

4. Identify opportunities for Kanban and determine Kanban locations

5. Illustrate these improvement opportunities as Kaizen bursts on future state VSM

6. Establish scheduling methods

7. Calculate performance data

A typical VSM shows the production data, process flow, material flow, and information flow in a system. An example of VSM is given in Figure 2.3.

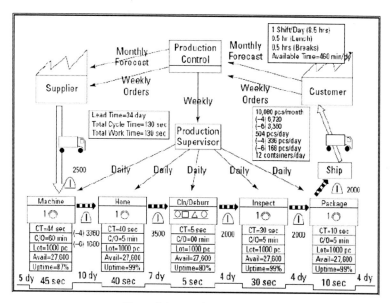

Figure 2.1   A Sample of VSM

Some basic and standard icons of VSM are given in Appendix B. From Figure 2.1, it can be seen that a VSM contains essential, descriptive process information. Generally, a VSM contains the following (Sayer and Williams, 2007):

1. Process Steps: The VSM shows the process steps in the value stream, including both value-added (VA) and non-value-added (NVA).

2. Inventory: The VSM highlights storage and the amount and movement of work-in-progress within the process.

3. Information flow: All supporting information required by the process is shown on the VSM. This can include orders, schedules, specifications, kanban signals (a kanban is a signal to replenish inventory in a pull system), shipping information, and more.

4. Box score: A VSM includes a summary of the key operational metrics of the process. At a minimum, this includes a summary of the total lead time of a process. The summary may also include such information as distance travelled, parts per shift, scrap, pieces produced per labour hour, changeover time, inventory turns, uptime, downtime, etc.

5. Lead time: Along the bottom of the VSM is the current lead time performance of the value stream. Lead time is the amount of time that one piece takes to flow completely through the process. The time is divided into value-added and non-value-added portions.

6. Takt time: A box in the upper-right-hand corner of the VSM shows the customer demand rate or takt time. This rate is determined by the customer demand and production time available. Ideally, all steps in the value stream should then produce to this rate.

Sayer and Williams (2007) highlighted the major differences between VSM and other process diagrams. These are listed down below:

❖ VSM always has the customer's perspective and is focused on delivering to the customer's expectations, wants, and needs.

❖ VSM, in a single view, provides a complete, fact-based; time-series representation of the stream of activities — from beginning to end — required delivering a product or service to the customer.

❖ VSM provides a common language and common view to analyze the value stream.

❖ VSM shows how the information flows to trigger and support those activities.

❖ VSM shows which activities are adding values and which are not.

In their publication, Sayer and Williams (2007) brought out an important point that the role and purpose of value stream mapping is not about conducting huge efforts, big projects, and long implementation programs. Value stream mapping is a concise effort, performed in a short time-span of just a few days. The major purpose of value stream mapping is to create a fundamental for Kaizen (Japanese: continuous improvements).

2.3   Principles of Lean Implementation

Womack and Jones (1996) stated that lean thinking can be summarized into five principles:

Step 1: Define value from the customer perspective. Value can only be defined by the ultimate customer and it is only meaningful when expressed in terms of a specific product (a finished good or a service, and often both at once) which meets the customer's needs at a specific price at a specific time. Lean technique and ideas assume that the only way to produce a product is from the customers' perspectives.

Step 2: Identify the value stream for each product: Value stream can be defined as the set of all specific actions required to bring a specific product through the three critical tasks of any business:

The problem-solving task: the process of running from concept through detailed design and engineering to the product launch.

The information task: the process of running from an order being placed through detailed scheduling to delivery.

The physical transformation task: starts at the raw material and ends with a finished product in the hands of the customers.

In this step, the entire value stream is identified for each part or product by using lean tools to analyze the types of activities that are considered waste:

❖ Process or steps that create value to the product.

❖ Process or steps that do not create value to the product but are unavoidable; referred to as type I waste.

❖ Additional process or steps that do not create value and are avoidable; referred to as type II waste.

Step 3: Make the value flow: After the process elimination steps have been completed, the next phase is to make the journey to value-added activities.

Step 4: Pull: Let the customer pull value from the producer (kanban).

Step 5: Perfection: The goal of the final stage is to pursue and strive for perfection to improve overall performance. Perfection is an endless process of reducing time, space, cost, and error.

2.4   The Benefits of being "LEAN"

In non-process industries (Figure 2.2), such as the automotive industry, the benefits of being lean are well documented (Melton, 2005):

1. Decreased lead times for customers;

2. Reduced inventories for manufacturers;

3. Improved knowledge management;

4. More robust processes (as measured by less errors and therefore less rework).

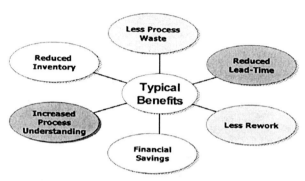

Figure2.2   The Benefits of 'Lean' (Melton, 2005)

Over the last 10-15 years, researchers have brought lean manufacturing practices to the US with astounding success. Companies using these practices report productivity improvements in the triple digits, defect rates falling by orders of magnitude, significantly shortened order-to-delivery times, increased customer satisfaction, greatly reduced employee turnover, and more. The concepts are quite literally revolutionizing manufacturing in the US. (Sobok. 2004).

The key to lean manufacturing is to compress the cycle time between receipt of an order and receipt of the payment. Compressing cycle time yields greater productivity, shorter delivery times, lower costs, improved quality, and increased customer satisfaction. There will be no overproduction in a lean manufacturing system, because overproduction is a result of "pushing" an order through the factory rather than pulling the desired quantity from the finished goods stores.

If one is looking for non-leanness or wastes, it will usually be found in the form of long lead times, or defective products. In a lean company, good products are produced when needed and in the form and quantity needed.

The origins of lean manufacturing can be traced back to the early 1900s in the US. Henry Ford introduced a new manufacturing system - mass production. Ford's philosophy was to build a small, strong and simple car at the lowest cost. Ford tried to reduce the prize of the product by shortening the production cycle. The key elements of the Ford manufacturing system were conveyors, division of labor, and an integrated supply chain (Imai, 1986).

The Toyota production system evolved from Ford manufacturing system. Managers and employees learned to question the need for every work sequence, every piece of in-process inventory, and every second that people, material and machines are idle. As a result, not only did production increase, but quality also increased when people learned to identify and eliminate waste (Ohno, 1988 and Monden, 1993).

2.5   Supply Chain Management

Supply chain management (SCM) is an advance that has evolved out of the combination of these considerations. SCM is known as the combination of key business procedures from last part user through original providers that present products, services, and information and therefore adds value for customers. SCM is an increasingly useful procedures pattern for increasing overall organizational competitiveness. (A. Gunasekaran, E.W.T. Ngai, 2004)

A latest study of supply chain linked executives found that 92% of those studied were scheduling to execute one or more supply chain ideas in 1999 (Bradley, 1999). SCM is based on the combination of all behaviors that increase the value to customers, staring from designing the product to delivery of it. Supply chain management is a point of approaches applied to efficiently combine suppliers, manufacturers, warehouses, and stores, so that products is created and spread at the exact amounts, to the precise positions, and at the correct time, to reduce system broad cost in order to pleasing service stage conditions. (A. Gunasekaran , E.W.T. Ngai, 2004)

Today's companies are in contest for civilizing their organizational competitiveness in order to participate in the latest century worldwide marketplace. This market is electronically linked and active in nature. Hence, companies are demanding to improve their nimbleness stage with the purpose of being supple and responsive to face to the changing marketplace requirements. In an attempt to reach this, many companies have spread out their value-adding actions by outsourcing and increasing virtual enterprise.

2.6   Supply Chain Management Information Systems

The acronym SCMIS will be used throughout the study to refer to Supply Chain Management Information Systems. SCMIS are progressively important to the prosperity of many companies (Chopra and Meindl 2001; Kumar 2001), but have received inadequate consideration in experiential information system study (Subramani 2004). Few researches have investigated the profits and abilities of various kinds of SCMIS such as EDI (Lee et al. 1999; Mukhopadhyay et al. 1995), electronic market (Dagenais and Gautschi 2002; Kaplan and Sawhney 2000), or extended enterprise resource scheduling (Green 2001) systems.

Nevertheless, there are few experimentally extracted models appropriate for examining the scope of SCMIS options. Similarly, companies face complicated and dangerous determination analyzing and choosing a suitable SCMIS solution or guaranteeing that their executed systems are organized with their competitive policies (Reddy and Reddy 2001). While the improved policies suitable model is operationalized particularly for SCMIS, the fundamental hypothesis and methodology could be accepted in approach studies hence, analyzing other kinds of information systems.

SCMIS has an extremely vital role in the ability of companies to decrease expenses and increase the accessibility of their supply chain. SCMIS are information systems used to organize information between indoor and outdoor customers, providers, distributors, and other members in a supply chain. Few researches have explored the benefits and potential of dissimilar SCMIS like Electronic Data Interchange, Electronic Marketplace, or Enterprise Resource Planning systems.

However, there are some experimentally form fit for evaluating the organizational skill held up by the ambit of SCMIS alternatives. Due to that, companies face complicated and perilous determinations appraising and picking up a suitable SCMIS solution or guaranteeing that their executed systems are organized with their business policies (Reddy and Reddy 2001). An organizational skill is the aptitude of an organization to attain its objectives by leveraging its different resources (Ulrich and Lake 1990).

Information system abilities are organizational abilities which are capable by information system. Correspondingly, SCMIS abilities are organizational abilities capable by the information system. Through the years, study on the estimation of information system has improved in abstraction from matching information system abilities with practical necessities (Lucas 1981), to desired formation (Allen and Boynton 1991), to competitive policies (Henderson et al. 1996).

Despite politic alignment has received important concentration in current researches of generally information systems policies (Kearns and Lederer 2001; Reich and Benbasat 2000; Sabherwal and Chan 2001), models have not yet been advanced to an adequately itemized stage to interrogate the organizational abilities allowed by peculiar kinds of information system, like SCMIS

2.7   Supply Chain Management and Lean Production

The system of interconnected businesses used to push a product from supplier to consumer is defined as a supply chain. SCM focuses on managing the supply chain in an effort to improve the quality and time it requires to manufacture a product. In addition to implementing SCM, a helpful lean production practice called "Just-in-time "can be used to remove any waste present along the supply chain. The marriage of lean production and SCM creates lean SCM, which provides a much leaner and more economical supply chain for the product to flow through.

SCM and lean production much uncertainty about what SCM entails is present in today's society. Many people treat SCM as being synonymous with logistics, which is the management of the flow of goods from the origin to the consumers (Lambert, 2008). However, supply chain management encompasses much more than the purchasing or management of goods to the consumer. SCM, as defined by Lambert (2008), is the management of relationships across the supply chain, which includes a network of interconnected businesses involved in providing a product or service to the consumer. The management of the relationships between businesses on the supply chain is significant to ensure successful and efficient processes are used in providing products and goods to the customer.

2.8   Information system in lean production

Manufacturers are torn between two "opposing" camps. In one corner is lean manufacturing. In the opposite camp is computer-based planning and control systems. The explosive growth of e-business is forcing many companies to revisit where they stand in this apparent conflict. Indeed, some wonder reconciliation is possible that would combine the best of both worlds. This new force is referred to here as e-Lean.

Spawned by the Toyota Production System, the classic lean movement has gained tremendous momentum and respect over the last decade. However, it has been almost anti-information systems (IS) in its stance.

- For Lean aficionados, less is best. This means:
- Less inventory
- Less material movement
- Less floor space
- Less variability
- Fewer steps, options and choices in work.

For those in the computer camp, on the other hand, more is best. This includes:

- More information
- More flexibility
- More functions and features
- More comprehensive business processes
- Faster, more frequent decision making made by more people.

Hence, it is no surprise that these two camps clash. However, it is a mistake for a manufacturer to pursue one of these strategies totally to the exclusion of the other. Indeed, tomorrow's most successful manufacturers will learn how to artfully blend the best of both worlds to create the next-generation powerhouses.

Many in the lean camp look at information systems (IS) almost with disdain. In their eyes, IS introduces far too many complications and extraneous tasks. Almost all would become moot if the manufacturing processes were set up right in the first place, they believe. For synchronizing material flow, for instance, no fancy computer control systems are needed. An empty container is sufficient to alert a worker to make more parts.

The sheer presence of considerable automation and computer systems likewise can drain manufacturing resources from a firm's core mission: producing vehicles and parts. The most high-tech-ish plants (e.g., Mitsubishi's Illinois plant) often spend inordinate amounts of time struggling with their systems and fixing wayward robots, not building product.

Even when the systems work flawlessly, computer systems can encourage band-aid solutions. For instance, a computer savvy worker may try clever work around instead of fixing a manufacturing problem (e.g., shoddy equipment maintenance or high worker absenteeism) at its source. In planning and execution, the lean camp also argues that computer-based planning and control dangerously removes control from the plant and over centralizes it. This has been especially true for traditional Material Resource Planning (MRP) and Enterprise Resource Planning (ERP) systems. Using these planning systems can lead to a major disconnect between reality on the plant floor and computer-generated schedules, inventory counts, and the like.

2.9   Information systems for lean

It is the information age without any doubt. This is because information is available to everyone faster and easier. But this doesn't mean all this information is required or useful. We still have to use this information effectively and efficiently in order to create value to the system. Lean manufacturing concepts can be used in streamlining your information flow.

A lean information system is essential for the success of any lean manufacturing system. Without this synchronization it is impossible to have a good.

lean manufacturing facility. Most of the organizations have very poor information systems. For example many organizations use email in day to day communication.

They use many contacts on their computer contact lists to keep people informed. But in many instances these people are not directly involved in the process and they do not need this piece of information. Think for a moment. How many valuable man hours are wasted in this? This kind of information handling can also lead to confusions.

Every waste mentioned in manufacturing context can be identified in the context of communication as well. Overloading of information corresponds to over production. Waiting is common for both manufacturing and communication. Avoiding all these wastes can make your information system lean.

2.10   Information system for lean production issues in supply chain

During the last two decades, managements have perceived outstanding global changes due to advance in technology, globalization of markets and new situation of political economy. In these times firms should reduce costs and offer products or services with reasonable and acceptable price to their customer for increasing the competitive ability and their profitability.

Lean Supply Chain (LSC) approach specifies the importance of reducing the variation and improving the flow which are the results of decrease in inventory and safety stock. Actually we can say this approach concentrates on the internal process of supply chain for the sake of making capability.

Modeling of the supply chain arrangement problem is considered in the information system progress structure (Fig. 2.3) in order that uphold the integration of decision making model factors and the other components of the information system. The process of creating a model describing the problem to be solved provides additional information systems' progress procedures with primary input information, like common supply chain explanation and modeling purpose (Charu Chandra and Jānis Grabis, 2007).

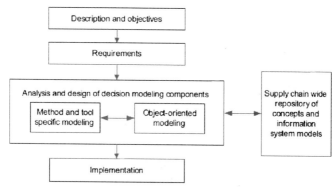

Figure 2.3   Interactions between information systems development and decision modeling (Charu Chandra and Jānis Grabis, 2007).

Methods, like utilizes case modeling can be utilized to identify necessities. For the supply chain arrangement problem, important necessities for the information system are implementation of decision-making procedures and integration of decision-making consequences through the supply chain information currents (Charu Chandra and Jānis Grabis, 2007).

2.11   Conclusion

This chapter highlighted the literature review a lean production, value stream mapping, principle of lean implementation, benefit of lean, supply chain management, supply chain management information system, supply chain management and lean production, information system in lean production, information system of lean and information system for lean production issues in supply chain.

Next chapter provides the methodology of the current study

CHAPTER 3

METHODOLOGY

3.1 Introduction

This chapter presents the details of the methodology used to develop the current state of value stream mapping (VSM) and used information model to show relationship for each state with other one in the manufacturing flow. The VSM as a lean manufacturing improvement tool will be described to enhance the understanding on how the study will be conducted.

3.2 Flow Chart of Research Methodology

The goal of this project is to assess the lean manufacturing. Then, according to the current VSM further improvements and find waste and try to eliminate of them. So, the project was started by searching for a manufactures company which that is appropriate to conduct this research.

After visiting several companies and evaluating their line production, Company A was selected to be that typical company. Then, several meetings were arranged to talk with the company manager and owner about selecting one production line that was running in the company at that time.

Finally, it was decided to get data to draw the current VSM and evaluate first view of the manufacturing with much waste in line production. Then, the available data about manufacturing such as type and amount and cost of raw materials used, amount of inventory, product manufacturing processes, costs of manufacturing the product, time study, transportation, some information about supplier and order processing, amount of operation and etc. were gathered.

The next step of the methodology was to group or categorize of the collected data according to the sub element of lean manufacturing such as, resource, cost, technology, process, information flow, raw material and time study etc. The grouping of the criteria or variable was based on the potential impact categories at which the criteria could be affected (Hemdi, A.R. *et al.*, 2010).refers the flow chart research methodology in Figure 3.1.

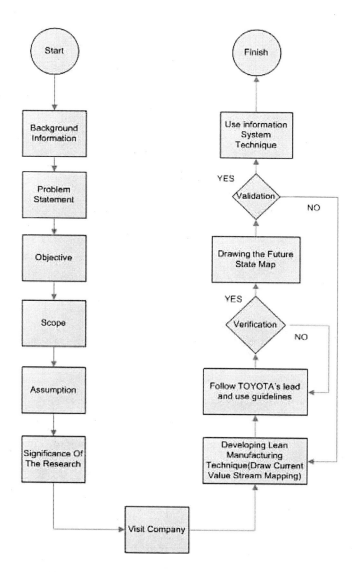

Figure 3.1   Flow Chart of Research Methodology

3.3   Value Stream Mapping Technique

Value stream mapping (VSM) is a technique that was originally developed by Toyota and then popularized by the book, Learning to See (The Lean Enterprise Institute, 1998), by Rother and Shook. VSM is used to find waste in the value stream of a product. Once identified, one can work to eliminate it. The purpose of VSM is to process improvement at the system level.

The research process flow for this research is presented in Figure 3.2 (DNREC, 2005). In the beginning stage researcher carefully narrowed the study by building the scope of the value stream. Researcher determined the value stream aspects that will be improved. For this study, researcher decided to conduct a study in manufacturing process and focus in trimming process in the company.

Next stage involved the preparation of current state drawing to have a clear understanding about the manufacturing flow. This stage is important because the researcher has to understand processes that currently operate in the plant. This is actually the foundation that leads to the future state creation.

The next approach followed by the future state drawing. At this stage researcher designed a lean flow that suggested an improvement to the current state drawing. Future state drawing would lead researcher to come with implementation plan that help to improve the current situation.

In the implementation plan, a detailed development was created. This implementation plan developed to support the objective of the research. All actions taken before lead to the implementation of the improved plan and that was the ultimate goal of the VSM.

As for the last stage, researcher used information system technique to develop the model and find relationship of each step with one another and how could control or flow be monitored. However, the researcher did not implement the improved plan in the company due to the limitation of the time. Management and approval except more suggestion as the improvement plan that could benefit the company.

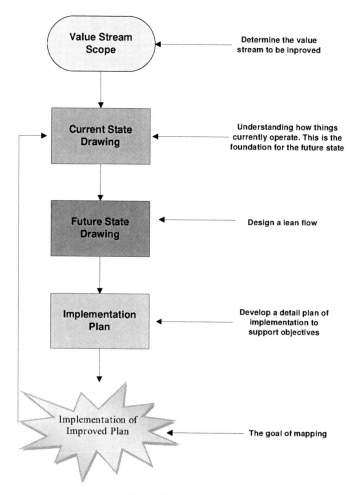

Figure 3.2   Flow Chart for using VSM

3.4   Process Flow

VSM shows the process in a normal flow format. However, in addition to the information normally found on a process flow diagram, VSM show the information flow necessary to plan and meet the customer's normal demands.

Other process information includes cycle times, inventories held, changeover times, staffing and modes of transportation, to name just a few. The typical VSM is called a "stock to dock" or "door to door" value stream map since it normally covers the information and process flow for the value stream at your facility. VSMs can be broader and cover any part or the entire value stream. However, functional responsibility often precludes the ability to take actions on these larger value stream maps. If actions are not the result of the value stream map, then the map has lost most of its effectiveness.

The key benefit to VSM is that it focuses on the entire value stream to find system wastes and tries to avoid the pitfall of optimizing some local situation at the expense of the overall optimization of the entire value stream. The strength of VSM may also be its weakness. It is not uncommon to find large wastes in cells, for example, which are not detailed on VSMs. If this is the case, large wastes can go unnoticed. This is a problem to those who only use VSMs in their battle to find and eliminate waste. VSM is only one tool in the battle for waste reduction, and to truly attack wastes, many tools are required.

VSMs detail information and the two key metrics highlighted are value added work and production lead time.

This part covers the implementation of some aspects of lean manufacturing in the company. The main focus for this section describes the current state VSM and problem identification. The analysis attempted to identify wastes in the system which is the non value added to the process flow. The improvement of the activities is one of the main focuses.

3.5   Data Collection

The whole process monitored and controlled by researcher. Researcher had combined different ways to collect the data. The results and conclusions were mainly derived from the exhaustive observation of each one of the application process phases. Interviews were carried out with members of the lean implementation team.

Researcher also followed the guidelines provided by the research methodology. Researcher exhaustively monitored the development of the VSM application in the company to analyze how effective the technique was and analyzed the keys to its correct application that lead to the suggestion implementation plan. In this study, researcher has concentrated on the manufacturing process and the main focus is on the trimming process.

Follow was the approaches done in order to do the collection.

- ❖ Studied the flow of manufacturing processes in detail.

- ❖ Identified wastes that should be removed.

- ❖ Made consideration of whether the process can be rearranged in a more efficient sequence.

- ❖ Made consideration of a better flow pattern, involving different flow layout or transport routing.

- ❖ Made consideration of whether everything that is being done at each stage is really necessary and identified what would happen if extra task were removed.

3.6 Current State Value Stream Mapping Analysis

The objective of current VSM is to create a picture of how products flow through the value stream which is from raw materials to the customer's end product. For this project, Figure 3.3 shows the current state map that demonstrates the work processes as they currently exist in the company. It is important to understand the needs for changes and also where an opportunities lie.

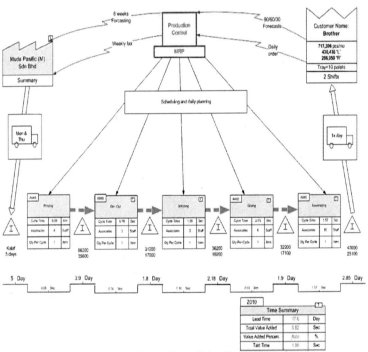

Figure 3.3   Current State Value Stream Mapping

3.7   Takt Time

Takt Time is a one of the key principles in a Lean Enterprise. Takt Time sets the 'beat' of the organization in synch with customer demand. As one of the three elements of "Just in Time" (along with one-piece flow and downstream pull) Takt Time balances the workload of various resources and identifies bottlenecks. Takt Time is a simple concept, yet counter-intuitive, and often confused with cycle time or machine speed. In order for manufacturing cells and assembly lines to be designed and built Lean, a thorough understanding of Takt Time is required. Several common misconceptions need to be dispelled, as well as introduce a rule of thumb of Takt Time for production line design.

Takt Time Formulate

Takt Time comes from a German word 'takt' meaning rhythm or beat. It is a term often associated with the takt the conductor sets so that the orchestra plays in unison. Takt Time is used to match the pace of work to the average pace of customer demand. Takt is not a number that can be measured, and is not to be mistaken with Cycle Time, which is the time it takes to complete one task. Cycle Time may be less than, more than, or equal to Takt Time.

You can never measure Takt Time with a stop watch. You must calculate it. The formula for Takt Time is:

$$Takt\ Time = \frac{Net\ Available\ Time\ per\ day}{Customer\ Demand\ per\ Day}$$

Takt Time is expressed as "seconds per piece", indicating that customers are buying a product once every so many seconds. Takt Time is not expressed as "pieces per second". By pacing production to this rate of customer demand, Lean Manufacturing seeks to minimize was and ensure on-time at a low cost.

3.8   Symbols for Value Stream Mapping

Value Stream Mapping symbols is not standardized and there are many variations. The most common symbols may be referred to in appendix A for specialized applications.

Table 3.1 : VSM Process Symbols

Customer/Supplier	This icon represents the Supplier when in the upper left, the usual starting point for material flow. The customer is represented when placed in the upper right, the usual end point for material flow.
Process Dedicated Process	This icon is a process, operation, machine or department, through which material flows. Typically, to avoid unwieldy mapping of every single processing step, it represents one department with a continuous, internal fixed flow path. In the case of assembly with several connected workstations, even if some WIP inventory accumulates between machines (or stations), the entire line would show as a single box. If there are separate operations, where one is disconnected from the next, inventory between and batch transfers, then use multiple boxes.
Process Shared Process	This is a process operation, department or work center that other value stream families share. Estimate the number of operators required for the Value Stream being mapped, not the number of operators required for processing all products.
C/T= C/O= Batch= Avail= Data Box	This icon goes under other icons that have significant information/data required for analyzing and observing the system. Typical information placed in a Data Box underneath FACTORY icons is the frequency of shipping during any shift, material handling information, transfer batch size, demand quantity per period, etc. Typical information in a Data Box underneath MANUFACTURING PROCESS icons: · C/T (Cycle Time) - time (in seconds) that elapses between one part coming off the process to the next part coming off, · C/O (Changeover Time) - time to switch from producing one product on the process to another · Uptime- percentage time that the machine is available for processing · EPE (a measure of production rate/s) -

	Acronym stands for "Every Part Every___". · Number of operators - use OPERATOR icon inside process boxes · Number of product variations · Available Capacity · Scrap rate · Transfer batch size (based on process batch size and material transfer rate)
Workcell	This symbol indicates that multiple processes are integrated in a manufacturing work cell. Such cells usually process a limited family of similar products or a single product. Product moves from process step to process step in small batches or single pieces.

Table 3.2 : VSM Material Symbols

Inventory	These icons show inventory between two processes. While mapping the current state, the amount of inventory can be approximated by a quick count, and that amount is noted beneath the triangle. If there is more than one inventory accumulation, use an icon for each. This icon also represents storage for raw materials and finished goods.
Shipments	This icon represents movement of raw materials from suppliers to the Receiving dock/s of the factory. Or, the movement of finished goods from the Shipping dock/s of the factory to the customers
Push Arrow	This icon represents the "pushing" of material from one process to the next process. Push means that a process produces something regardless of the immediate needs of the downstream process.
Supermarket	This is an inventory "supermarket" (kanban stockpoint). Like a supermarket, a small inventory is available and one or more downstream customers come to the supermarket to pick out what they need. The upstream work center then replenishes stocks as required. When continuous flow is impractical, and the upstream process must operate in batch mode, a supermarket reduces overproduction and limits total inventory.
Material Pull	Supermarkets connect to downstream processes with this "Pull" icon that indicates physical removal.
MAX=XX **FIFO Lane**	First-In-First-Out inventory. Use this icon when processes are connected with a FIFO system that limits input. An accumulating roller conveyor is an example. Record the maximum possible inventory.
Safety Stock	This icon represents an inventory "hedge" (or safety stock) against problems such as downtime, to protect the system against sudden fluctuations in customer orders or system failures. Notice that the icon is closed on all sides. It is intended as a temporary, not a permanent storage of stock; thus; there should be a clearly-stated

External Shipment	Shipments from suppliers or to customers using external transport.

Table 3.3 : VSM Information Symbols

Symbol	Description
Production Control **Production Control**	This box represents a central production scheduling or control department, person or operation.
Daily **Manual Info**	A straight, thin arrow shows general flow of information from memos, reports, or conversation. Frequency and other notes may be relevant.
Monthly **Electronic Info**	This wiggle arrow represents electronic flow such as electronic data interchange (EDI), the Internet, Intranets, LANs (local area network), WANs (wide area network). You may indicate the frequency of information/data interchange, the type of media used ex. fax, phone, etc. and the type of data exchanged.
Production Kanban	This icon triggers production of a pre-defined number of parts. It signals a supplying process to provide parts to a downstream process.
Withdrawal Kanban	This icon represents a card or device that instructs a material handler to transfer parts from a supermarket to the receiving process. The material handler (or operator) goes to the supermarket and withdraws the necessary items.
Signal Kanban	This icon is used whenever the on-hand inventory levels in the supermarket between two processes drops to a trigger or minimum point. When a Triangle Kanban arrives at a supplying process, it signals a changeover and production of a predetermined batch size of the part noted on the Kanban. It is also referred as "one-per-batch" kanban.
Kanban Post	A location where kanban signals reside for pickup. Often used with two-card systems to exchange withdrawal and production kanban.
Sequenced Pull	This icon represents a pull system that gives instruction to subassembly processes to produce a predetermined type and quantity of product, typically one unit, without using a supermarket.

XOXO **Load Leveling**	This icon is a tool to batch kanbans in order to level the production volume and mix over a period of time
MRP/ERP	Scheduling using MRP/ERP or other centralized systems.
Go See	Gathering of information through visual means.
Verbal Information	This icon represents verbal or personal information flow.

Table 3.4 : VSM General Symbols

Kaizen Burst	These icons are used to highlight improvement needs and plan kaizen workshops at specific processes that are critical to achieving the Future State Map of the value stream.
Operator	This icon represents an operator. It shows the number of operators required to process the VSM family at a particular workstation.
Other Information **Other**	Other useful or potentially useful information.
Timeline	The timeline shows value added times (Cycle Times) and non-value added (wait) times. Use this to calculate Lead Time and Total Cycle Time.

3.9   Information System Technique

Information systems are the means by which people and organizations, utilizing technologies, gather, process, store, use and disseminate information

3.9.1   Theoretical Underpinnings of Information Systems

The notions of wholeness, boundary, environment, emergence, communication, co-ordination and control are fundamental to the understanding of IS. IS draws primarily on systems theory, which provides an intellectual foundation and a basis for studying the enterprise as a complex adaptive system. The practice of IS necessitates the integration of systems theory and theories from other disciplines relevant to the range of application domains.

3.9.2   Data Information and Knowledge Management

Understanding of how data, information and knowledge can be modeled, stored, managed, processed and disseminated by computer systems. Knowledge of techniques and technologies used to organize data and information and enable their effective use by individuals, groups and organizations.

3.9.3   Information in organizational decision making

Providing relevant information for decision making is a primary function of an information system. Creating and utilizing information systems for effective decision making requires the system designer to understand what constitutes pertinent information and the context in which decisions are made.

3.9.4   Drawing Data Flow Diagram (DFD)

A data flow diagram (DFD) is a graphical representation of the "flow" of data through an information system, modeling its process aspects. Often they are a preliminary step used to create an overview of the system which can later be elaborated. (Bruza, P. D., Van der Weide, Th. P)DFDs can also be used for the visualization of data processing (structured design).

A DFD shows what kinds of data will be input to and output from the system, where the data will come from and go to, and where the data will be stored. It does not show information about the timing of processes, or information about whether processes will operate in sequence or in parallel.

Data flow diagram also approach can be used as tool to identify a system (John Azzolini, 2000). Therefore data flow diagram is a graphical representation of the flow of data through an information system. It can also be used for the visualization of data processing. Figure 3.4 shows the data flow diagram of the company in improved condition.

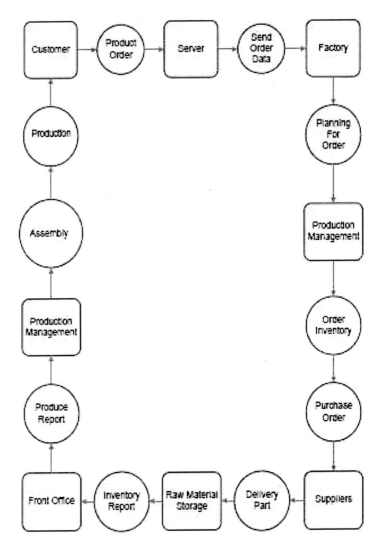

Figure 3.4   Data Flow of Diagram

3.9.5 Defines Integration Definition for Information Modeling (IDEF1X)

IDEF1X is a method for designing relational databases with a syntax designed to support the semantic constructs necessary in developing a conceptual schema. A conceptual schema is a single integrated definition of the enterprise data that is unbiased toward any single application and independent of its access and physical storage. Because it is a design method, IDEF1X is not particularly suited to serve as an AS-IS analysis tool, although it is often used in that capacity as an alternative to IDEF1. IDEF1X is most useful for logical database design after the information requirements are known and the decision to implement a relational database has been made. Hence, the IDEF1X system perspective is focused on the actual data elements in a relational database. If the target system is not a relational system, for example, an object-oriented system, IDEF1X is not the best method.

There are several reasons why IDEF1X is not well-suited for non-relational system implementations. IDEF1X requires, for example, that the modeler designate a key class to distinguish one entity from another, whereas object-oriented systems do not require keys to individuate one object from another. Further, in those situations where more than one attribute or set of attributes will serve equally well for individuating IDEF1X entities, the modeler must designate one as the primary key and list all others as alternate keys. Explicit foreign key labeling is also required. The resulting logical design IDEF1X models are intended to be used by the programmers who take the blueprint for the logical database design and implement that design.

However, the IDEF1X modeling language is sufficiently similar to IDEF1 in that models generated from the IDEF1 information requirements can be reviewed and understood by the ultimate users of the proposed system.

IDEF1X is a data modeling language for the developing relational database. It is used to produce graphical information model which represents the structure and

relate of information within an environment or system. Current IDEF1X of the
company can be found in Figure 3.5.

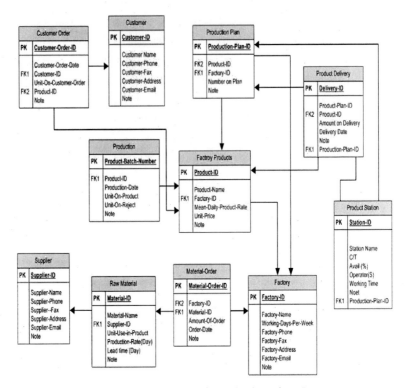

Figure 3.5   IDEF1X for Packaging Printer Box

3.9.3   Current Data Relationship

3.9.3.1   Connection Relationships

Connection relationships (solid or dashed lines with filled circles at one or both ends) denote how entities (sets of data instances) relate to one another. The connection relationships are always between exactly two entities. The connection relationship beginning at the independent parent entity and ending at the dependent child entity is labeled with a verb phrase describing the relationship. Each connection relationship has an associated cardinality. The cardinality specifies the number of instances of the dependent entity that are related to an instance of the independent entity.

3.9.3.2   Categorization Relationships

Categorization relationships allow the modeler to define the category of an entity. An entity can belong to only one category. For instance, there could be an entity CAR that is the generic entity in a category showing different types of cars. Each category entity must have the same primary key as CAR. Also, there must be a way of distinguishing between the category entities. The category entities are distinguished by a discriminator attribute which must have a different value for each category entity.

3.9.3.3   Attributes

Attributes are properties used to describe an entity. Attribute names are unique throughout an IDEF1X model, and the meaning of the names must be Every consistent. Each use has a range of meaningful values, and thus, the entity must be distinctly named. Each attribute is owned by exactly one entity. The attribute could be used in many places in a model, but would be owned by only one entity. Other occurrences of the social security number attributes would be inherited across relations.

Every attribute must have a value (No-Null Rule), and no attribute may have multiple values (No-Repeat Rule). Rules enforce creating proper models. In a situation where it seems that a rule cannot hold, the model is likely wrong

Figure 3.6 shows the relationship between supply chain members. This figure is created with Microsoft access 2007.

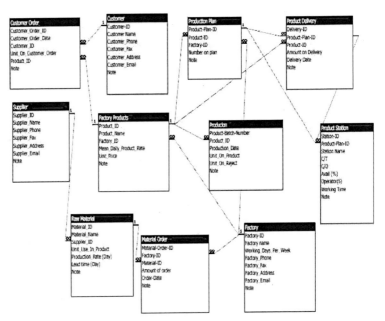

Figure 3.6   Relationships between Supply Chain Members

3.10 Conclusion

This chapter has presented the methodology in the research and how to conduct the value stream mapping in order to conduct improvement. The next chapter will present case study for better understanding of the manufacturing situation.

CHAPTER 4

CASE STUDY IN BOARD-PRINTING FACTORY

4.1   Introduction

This chapter presents a description of the company for this project. The company background and the overview of the company operation will be described in detail to enhance the understanding to the topic.

4.2   Company Background

Company A is a joint-venture company between a Malaysian company and a Japanese company. Company A commenced operations in October 1996 specializes in the manufacture of die-cut cartons made of special packaging products. This new range introduced by Company A aimed to replace The conventional cushioning materials such as synthetic foam and plastic which are not bio-degradable.the new products are highly favored in the market for their advantages of easy set up, high shock absorption and competitive prices.

Located in Johor Bahru, company A thus enjoys being the key local player in the manufacture of special die-cut cartons. In view of Malaysia's current export-oriented industrialization process and global emphasis on increased use of environmental-friendly products, the timely arrival of company A's new product range is expected to boost the local packaging industry and enhance Malaysia's export industry.

4.2.1   Company A Cushion

A recycle cushioning material made of corrugated paper. It is suitable for variety of applications ranging from heavy items to precision instruments. This cushion is easy to set up and provides high dimensional precision, making trouble-free operation in automatic packaging lines possible.

4.2.2   One Piece Box

One-piece carton developed in consideration of environmental protection. The box includes a cushioning structure that dispenses with polystyrene foam with the capability of being set up with a single action, the box have a considerable total merit including reduced storage, transportation cost and satisfactory product protection achieved by high cushioning performance.

4.3   Mills Division

Company A owns one of the largest Paper Mills in Malaysia in terms of capacity and offers a wide range of products to our domestic and overseas customers. They are the leading producer of various industrial grade paper comprising Test Liner, Corrugated Medium, Laminated Chip Board, Core Board, Grey Chip Board, Yellow Wrapping Paper, Inserting paper, Manila paper, MF Kraft.

Their mills have been accredited with ISO 9001 and OHSAS 18001 and their product qualities have gained industry-wide recognition and acceptance by their customers. The branding of "Board" has now won international recognition through their continuous presence in annual international trade fairs.

In line with their corporate mission, we continue to grow in tandem with the growth in market demand. We will be adding new machine capacities in 2010 to achieve a combined output of over 500,000 metric tons per annum.

Their mills are equipped with effluent treatment facilities to ensure that the mills effluent discharge would not cause harm to the environment. The mill was also equipped with a CHP plant to further save on energy consumption and to reduce carbon emission.

4.4   Packaging and Converting Division

Their corrugating and converting plants are strategically located throughout Peninsular Malaysia to provide packaging solutions to customers from various business sectors consisting of SMES to multinational companies. This geographical spread allows us to serve the customers better and faster.

With two paper mills in the Group, the Group's corrugated cartons plants are assured of uninterrupted supply of quality recycled industrial grade paper. With this Group synergy, the Group's cartons plants are able to provide quality value added products at competitive prices, reliable packaging solutions and on-time deliveries to its customers consistently.

They aim to be the preferred carton and packaging solution provider for all their customers and in line with the demand for more sophisticated packaging solutions, they are always on the lookout for innovative ways to meet customers' packaging requirement.

4.5   Marketing and Trading Division

The future growth of the Group's main business divisions requires the continual support of its existing customers and expansion of customers' base.

To serve their domestic and overseas customers efficiently, the Group has extensive marketing and sales offices throughout Malaysia as well as in Hong Kong, Australia and Singapore.

4.6   Products

Products by company A are subject to stringent quality control. Our laboratories are equipped with the best equipment to carry out all necessary quality tests procedures to ensure total customer satisfaction.

R & D work on product innovations is also carried out by their companies in order to offer products with better value added and packaging solutions to their customers.

4.7   Production Flow

As a business grows the scale of its operations, it often needs to change its method of production to allow greater production capacity.

A small business might use job or batch production to provide a personalized or distinctive product. However, if the product is intended for much larger, mass markets, then alternative methods of production may be required in order for the product to be produced efficiently. A key production method in these circumstances is flow production.

4.7.1   Printing Process Control

- Planning personnel shall forward "Daily Production Schedule" (MCP-PC-F06) and necessary documents such as "Master Card" (MCP-SA-F05) and "Work Order" (MCP-PC-F02) to production personnel before starting of printing process. Refer to MCP-PC-P03 (Production Planning and Material Control).

- Production personnel shall base on the order's requirements (including RoHS requirements) and request for .material from Store Department via "Stock Requisition" (MCP-ST-F01). Refer to MCP-ST-P02 (Receiving, Issuing and Delivery)

- At the same time, tooling is prepared. Tooling shall be controlled and preserve as per MCP-PD-W02 (Tooling Control).

- Machine set up and preparation of ink shall be made in accordance with MCP-PD-W01 (TCY 1 Operating Instruction) or MCP-PD-W08 (TCY 2 Operating Work Instruction)

- Parameter setting is as per MCP-PD-W03 (First Piece Sample Confirmation)

- First piece produced shall be verified by supervisor / leader before Mass Production. This is done by comparison between "Master Card" (MCP-SA-F05) and "Customer approved sample" / "Confirmed Production Sample"

- If the model is an initial run at the production, Customer approved sample shall be used for verification instead;

- For details of first piece confirmation, refer to MCP-PD- W03 (First Piece Sample Confirmation)

- If verification of the sample not satisfactory then re-set the machine and sample confirmation shall be conducted again.

- Daily production output shall be updated into "Work Order" (MCP-PC-F02) and recorded into "Printing Daily Production Report" (MCP-PD-F01)

- All WIPs after printing process shall be identified in accordance with MCP-PD-P02 (Identification and Traceability).

- By referring to the "Master Card" (MCP-SA-F05), the,; WIPs are then moved to the next production process (e.g. die cut, Stitching, gluing or assembly).

- Monitoring and measurement of the products are in accordance with MCP-QA-P0i (Inspection and Testing).Refer to Figure 4.1.

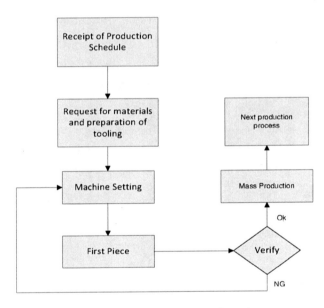

Figure 4.1   Printing Process Control

4.7.2   Die-Cutting Process Control

- Die cutting personnel to proceed with production by referring to their own "Daily Production Schedule" (MCP-PC-F06) and necessary documents such as "Master Card" (MCP-SA-F05) and "Work Order" (MCP-PC-F02).

- Production personnel shall base on the order's requirements (including RoHS requirements) and prepare the material with the Sales Order identification number at the production area. Refer to MCP-PD-P02 (Identification and Traceability).

- Raw materials are requested from Store via the issuance of "Stock Requisition" (MCP-ST-F01) if die cut is the first process in production.

- At the same time, tooling is prepared. Tooling shall be controlled and preserved as per MCP-PD-W02 (Tooling Control).

- Die Cut Machine set up shall be made in accordance with MCP-PD-W04 (AP16 Operating Work Instruction)

- Parameter setting is as per MCP-PD-W03 (First Piece Sample Confirmation).

- First piece produced shall be verified by supervisor / leader before Mass Production. This is done by comparison between "Master Card"(MCP-SA-F05).

- If the model is an initial run at the production, all dimensions shall be checked. Refer to MCP-QA-P01 (Inspection and Testing).

- For details of first piece confirmation, refer to MCP-PD- W03 (First Piece Sample Confirmation).

- If verification of the sample not satisfactory, then re-set the die cut machine and sample confirmation shall be conducted again

- Daily production output shall be updated into "Work Order" (MCP-PC-F02) and recorded into Die "Cutting Daily Production Report" (MCP-PD-F02).

- All WIPs after die cut process shall be identified in accordance with MCP-PD-P02 (Identification and Traceability).

- By referring to the "Master Card" (MCP-SA-F05), the WIPs are then moved to the next production process (e.g. stitching, gluing or assembly)

- Monitoring and measurement of the products are in accordance with MCP-QA-P01 (Inspection and Testing).Refer to Figure 4.2.

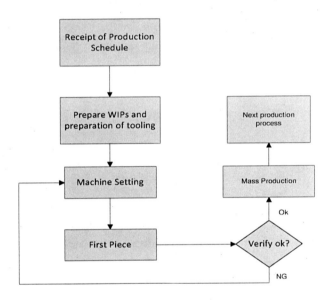

Figure 4.2   Die-Cutting Process Control

4.7.3   Stitching Process Control:

- Respective production section personnel shall proceed with production by referring to their respective "Daily Production Schedule" (MCP-PC-F06) and necessary documents such as "Master Card" (MCP-SA-F05) and "Work Order" (MCP-PC-F02).

- Production personnel shall base on the order's requirements (including RoHS requirements) and prepare the material (WIPs)' with the Sales Order identification number at the production area. Refer to MCP-PD-P02 (Identification and Traceability).

- Stitching machine set up and preparation of material shall be made in accordance with MCP-PD-W06 (Stitching Operating Work Instruction).

- Parameter setting is as per MCP-PD-W03 (First Piece Sample Confirmation).

- First piece produced shall be verified by Supervisor / Leader before Mass Production. This is done by comparison between "Master Card"(MCP-SA-F05).

- First piece confirmation of stitching shall be refers to MCP-PD-W03 (First Piece Sample Confirmation).

- If .verification of the sample not satisfactory, then re-set machine is required and sample confirmation shall be conducted again.

- Daily production output shall be updated into "Work Order" (MCP-PC-F02) and recorded into Stitching Daily Production Report (MCP-PD-F05).

- All WIPs/Finished goods after process shall be identified in accordance with MCP-PD-P02 (Identification and Traceability).

- By referring to "Master Card" (MCP-SA-F05), the WIPs are then moved to the next production process (if any) or to Store Department

- Finished goods shall be outgoing inspected before transfer to Store. Refer to Figure 4.3.

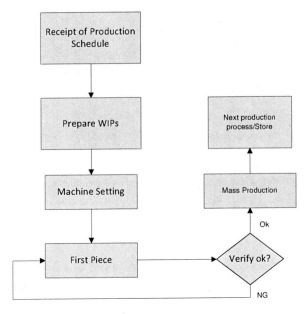

Figure 4.3   Stitching Process Control

4.7.4   Gluing and Assembly Process Control

- Respective production section personnel shall proceed with production by. Referring to their respective Daily Production Schedule (MCP-PC-F06) and necessary documents such as "Master Card" (MCP-SA-F05) and "Work Order" (MCP-PC-F02).

- Production personnel shall base on the order's requirements (including RoHS requirements) and prepare the material (WIPs) with the Sales Order identification number at the production area. Refer to MCP-PD-P02 (Identification and Traceability).

- Gluing Machine set up and preparation of glue shall be made accordance with MCP-PD-W05 (FG9N Operation Work Instruction) or MCP-PD-W07 (Glue Operating Work Instruction).

- Whereas, for Assembly process, no work instruction is required, only need to refer to the model "Master Card" (MCP-SA-F05).

- First piece produced shall be verified by supervisor / leader before Mass Production. This is done by comparison between "Master Card" (MCP-SA-F05).

- First piece of gluing shall be accordance with MCP-PD- W03 (First Piece Sample Confirmation).

- If verification of the sample not satisfactory, then resetting is required and sample confirmation shall be conducted again.

- Daily production output shall be updated into "Work Order" (MCP-PC-F02) and recorded into "Gluing Daily Production Report" (MCP-PD-F03), "Semi Glue Daily Production Report" (MCP-PD-F06) or "Assembly Daily Production Report" (MCP-PD-F04).

- All WIPs/Finished goods after process shall be identified in accordance with MCP-PD-P02 (Identification and Traceability).

- By referring to the "Master Card" (MCP-SA-F05), the WIPs are then moved to the next production process (if any) or to Store department.

- Finished goods shall be outgoing inspected before transfer to Store.

- Monitoring and measurement of the products are in accordance with MCP-QA-P01 (Inspection and Testing).Refer to Figure 4.4.

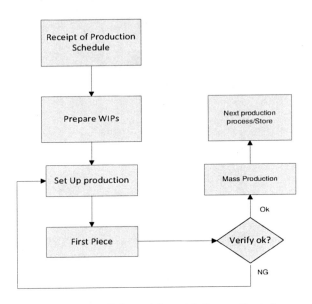

Figure 4.4   Gluing and Assembly Process Control

4.8   Production Layout

Plant Layout is the physical arrangement of equipment and facilities within a Plant. The Plant Layout can be indicated on a floor plan showing the distances between different features of the plant. Optimizing the Layout of a Plant can improve productivity, safety and quality of Products. Unnecessary efforts of materials handling can be avoided when the Plant Layout is optimized. This is valid for:

- Distances Material has to move

- Distances Equipment has to move

- Distances Operators have to move

- Types of Handling Equipment needed

- Energy required moving items against resistance

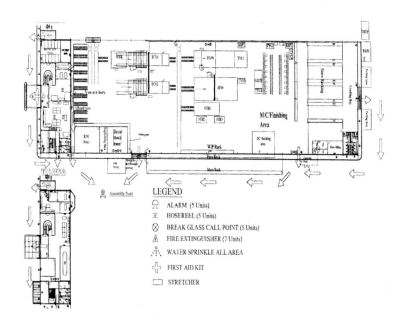

LEGEND

ꙮ ALARM (5 Units)

☰ HOSEREEL (5 Units)

⊗ BREAK GLASS CALL POINT (5 Units)

△ FIRE EXTINGUISHER (7 Units)

⸭ WATER SPRINKLE ALL AREA

✛ FIRST AID KIT

▭ STRETCHER

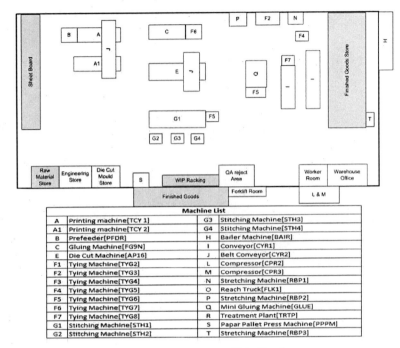

Figure 4.5 Layout and Machine list

	Machine List		
A	Printing machine[TCY 1]	G3	Stitching Machine[STH3]
A1	Printing machine[TCY 2]	G4	Stitching Machine[STH4]
B	Prefeeder[PFDR]	H	Bailer Machine[BAIR]
C	Gluing Machine[FG9N]	I	Conveyor[CYR1]
E	Die Cut Machine[AP16]	J	Belt Conveyor[CYR2]
F1	Tying Machine[TYG2]	L	Compressor[CPR2]
F2	Tying Machine[TYG3]	M	Compressor[CPR3]
F3	Tying Machine[TYG4]	N	Stretching Machine[RBP1]
F4	Tying Machine[TYG5]	O	Reach Truck[FLK1]
F5	Tying Machine[TYG6]	P	Stretching Machine[RBP2]
F6	Tying Machine[TYG7]	Q	Mini Gluing Machine[GLUE]
F7	Tying Machine[TYG8]	R	Treatment Plant[TRTP]
G1	Stitching Machine[STH1]	S	Papar Pallet Press Machine[PPPM]
G2	Stitching Machine[STH2]	T	Stretching Machine[RBP3]

4.9 Procedure

4.9.1 Responsibility

Head of Store department shall be responsible for the implementation and effectiveness of the procedure

4.9.2 Receiving and Issuance of Raw Materials

- Store personnel is responsible for the receiving of incoming materials and verify the quantity as well as items delivered are correct by referring to the

CPS's Receiving List or "Delivery Instruction" (MCP-PC- F04) against physical items sent and Supplier's Delivery Order (DO).

- Physical quantity shall update into "Delivery Instruction" (MCP-PC-F04) for monitoring purposes.

- Upon verify "OK", Store personnel shall update into CPS - Warehouse Receiving Log and inform Quality Assurance Department to perform incoming inspection. Refer to MCP-QA-PQ1 (Inspection and Testing)

- All incoming materials shall be preserved in accordance to Product Preservation and Aging Control (MCP-ST-P01)

- All requisition shall be made via issuance of "Stock Requisition Foim"(MCP-ST-F01)

- Updating into CPS is required to be done immediately upon issuance. Refer to Figure 4.6.

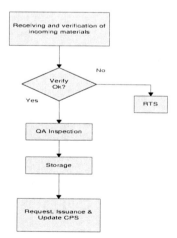

Figure 4.6   Receiving and Issuance of Raw Material

4.9.2.1 Return of Excess Material

Production personnel shall be responsible to return all excess material at the end of the day via "Stock Transfer Form" (MCP-ST-F04) or alter in "Stock Requisition" (MCP-ST-F01).

4.9.3   Delivery of Finished Goods

- All finished goods transferred from production shall be updated into "GPS - Floor Tracking System"

- Rework finished goods will be transferred back to Store via "Stock Transfer Form" (MCP-ST-F04)

- Sales/Store Personnel is to print out "CPS Sales Order" "Delivery Schedule" base on customer delivery7s requirement and forward to Store Personnel

- Any cancellation of "Delivery Order" or invoice shall be justified by Store Executive or Sales Executive on the "Delivery Order" copy before submit to Factory Manager for approval

- Cancelled copy of "Delivery Order" & "Invoice" shall be kept & file accordingly

- Store personnel to arrange lorry for delivery and prepare "Delivery Order" (MCP-ST-F02) & Invoices

- All deliveries made are updated into "Daily Delivery Schedule" (MCP-ST-F03).Refer to Figure 4.7.

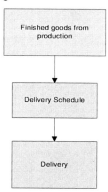

Figure 4.7   Deliveries of Finished Goods

4.9.4 Top Up Of Finished Goods at Store Area

All finished goods to be topped up for deliveries are required to follow as below:

- Storekeeper to confirm pallet slip with actual goods on model name & part number.

- Cover top of the remaining balances stock with stretch film.

- Paste top up sticker on top of remaining balance on the pallet

- Inform to QA on top up procedure to be performed. (Applicable to required customer only). The transaction shall be recorded using Stock Top Up Record (MCP-ST-F06).

- QA inspector to counter confirm on models and part number.

- Storekeeper to perform top up with the witness of QA inspector.

- Wrap the pallet with the stretch film and ready for delivery.

4.10   Conclusion

In this chapter researcher has described the information flow for the processes in the factory. The next chapter will presents the results and discussion that describes the methods used by the researcher to collect and analyze data.

CHAPTER 5

RESULTS AND DISCUSSION

5.1   Introduction

This chapter discussed the future state VSM structure. All the propose improvement activities are stated here. The propose improvement will be discuss based on the problem identified from the previous observation

5.2   The Future-State Map

The purpose of VSM is to highlight sources of waste and eliminate them by implementation of a future-state value stream that can become a reality within a short period of time. The goal is to build a chain of production where the individual processes are linked to their customers either by continuous flow or pull, and each process gets as close as possible to producing only what its customer(s) need when they need it.

Assuming that someone works at an existing facility with an existing product and process, some of the waste in a value stream will be the result of the product's design, the processing machinery already bought, and the remote location of some activities. These features of the current state probably can't be changed immediately. Unless a new- product is introduced, the first iteration of the future-state map should take product designs, process technologies, and plant locations as given and seek to remove as quickly as possible all sources of waste not caused by these features. (With the exception of minor purchases think, "What one can do with what one has?") Subsequent iterations can address the product design, technology, and location issues.

It is found that the most useful aid for helping people draw future-state maps is the following list of questions. As the future-state concepts, are developed answer the questions in roughly the following order. Based on the answers to these questions, mark the future-state ideas directly on the current-state map in red pencil. Once the future-state is worked out a future-state map can be drawn.

5.3   Drawing the Future-State Map

Perhaps the most striking things about Company A are the large amounts of inventory, the unconnected processes (each producing to its own schedule) pushing their output forward, and the long lead time in comparison to the short processing time. These can be sorted out accordingly a show below:

5.3.1   Company A Takt time

The takt time calculation starts with the available working time for one shift in A's assembly area, which is 28,800 seconds (8 hours). From this you subtract any non-working time, which is two 30-minute breaks per shift. The customer demand of 13,796 units per shift is then divided into the available working time to give a takt of 1.95 seconds.

$$Takt\ Time = \frac{Availble\ Work\ Time}{Customer\ Demand}$$ Equation (1)

	Time	Breaks & Rest Time
Available Work Time	8 hrs*60 Min*60 sec =28800	30 min*60 sec=1800
	28800-1800=**27000**	

	Total Production	Production in per Daily shifts
Customer Demand	717396/26 Days=27,592	27,592/2shifts=**13,796**

$$Takt\ Time = \frac{Availble\ Work\ Time}{Customer\ Demand} = \frac{27000}{13796} = 1.95$$

-   Working time to give a takt of 1.95 seconds.
-   Available Working Time: 28,800 - 1800 = 27,000 seconds per shift
-   Available Working Time 27,000 sec. + 13,796 units per shift Customer Demand

- A's Box Assembly Takt Time = 1.95 seconds

Company A may decide to cycle finishing faster than takt, if it is not possible to eliminate downtime problems. But the takt time is a reference number defined by the customer and cannot be changed by company A decision.

5.3.2 Finished-Goods Supermarket

At Company A, boxes are many paper (easy to store) parts that have only two varieties. The customer's demand rises and falls somewhat unpredictably, and company A is uncertain about the reliability of future- state changes to be made. So company A has opted to begin with a finished-goods supermarket and to move closer to "produce to shipping" in the future. For custom products we may not be able to create a supermarket of finished goods. See diagram at figure 5.1.

5.3.3 Building to a Supermarket

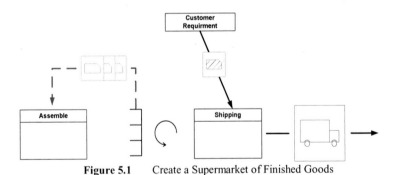

Figure 5.1   Create a Supermarket of Finished Goods

5.3.4 Building Directly to Shipping:

Figure 5.2 shows the production control scheduling assembles.

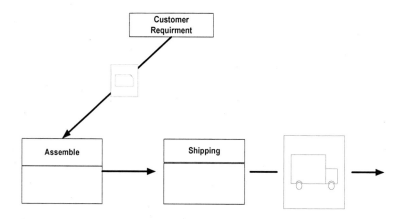

Figure 5.2   Production Control Scheduling Assemble

5.4 Introducing the Continuous Flow

The "operator-balance chart" summarizes the current total cycle times for each process. The printing operation cycles very quickly (0.08 second per part) and changes over to serve several product lines. So incorporating it into a continuous flow, which would mean slowing its cycle to near takt time and dedicating it to the Box product family, is not practical. That would result in a vastly underutilized printing and the need to buy another expensive printing machine for A's other product lines! It makes more sense to run A's printing as a batch operation and control its production with a supermarket- based pull system.

Examining the two assembly workstations, it is noticed that their cycle times are not too far apart and near the takt time as well. These workstations are also already dedicated to the Box product family, so continuous flow in assembly certainly is a possibility. The same is true for the two Die-Cut and Stitching workstations, where work could also pass directly from one step to the next in a continuous flow.

What prevents A from using continuous flow all the way from cutting through assembly, a condition with no inventory? The lean approach is to place these four processes immediately adjacent to each other, have the operators carry or pass parts from one process step to the next, and distribute the work elements so that each operator's work content is just below takt time.

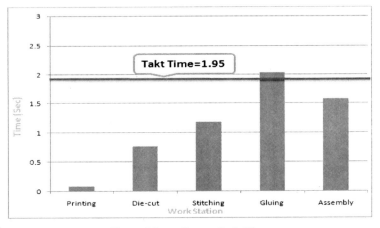

Figure 5.3   Current Cycle Time

Dividing the total die-cutting, stitching, gluing and assembling work content by the takt time (5.54 seconds divided by 1.95) reveals that 2.84 operators would be needed to run cutting, stitching and gluing in a continuous flow at takt.

The next option is to eliminate waste through process kaizen to bring the work content under the takt time ceiling. A kaizen target might be to reduce each operator's work content to 1.85 seconds or less (or ^ 5.55 seconds total work content.) If that fails, use of some overtime may be necessary.

To allow production to takt time and mix leveling, a pacemaker process should ideally incur little or no changeover time and change over very frequently. So the right-drive to left-drive die-cutting changeover times will need to be reduced from the current 15 minutes to a few seconds. Figure 5.4 shows the new takt time after balancing with three stations.

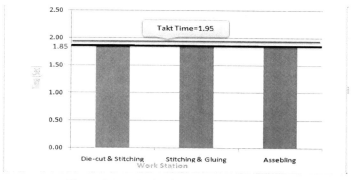

Figure 5.4   Cycle Time after Process Kaizen

5.5   First view of the Future state map

Figure 5.5 shows the first view of the future state map showing, Takt time, die-cutting, stitching and gluing cell and the finished-goods supermarket

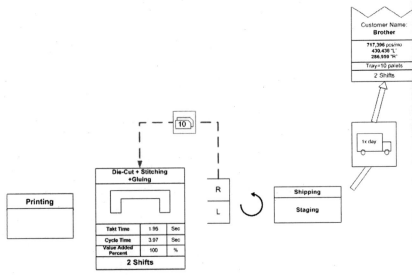

Figure 5.5   First View of the Future-State Map

This future-state map the three die-cutting, stitching and gluing process boxes have been combined into one process box to indicate the continuous flow. A small schematic sketch of a cell inside the process box also indicates the cellular manufacturing idea.

5.6   Use Supermarket Pull Systems?

Company A has decided to produce printing box to a finished-goods supermarket. Two additional supermarkets—one for Printing parts and one for pallet are necessary to complete Company A's in-plant value stream for printing boxes (packaging).

5.6.1 Printed Parts

Ideally it is possible to introduce a tiny printing machine dedicated to printing boxes called a "right-sized tool" and incorporate this mini-print into the die-cutting, stitching and gluing continuous flow. Unfortunately, this is not possible in the immediate future because machinery of this type does not yet exist. So it is necessary to set up a supermarket and a use withdrawal from that supermarket (pull) to control printing's production of right-drive and left-drive parts.

Pull system design begins with customer requirements, and printing's customer here is the die-cut, stitch and glue cell. The cell currently requires approximately 16,555 RI 1 and 11,036 LI printed parts per day. Containers for the printed parts should be sized to allow close- to-the-fingertips placement in the cell not primarily for the convenience of the printing or material handling departments! Small containers allow Company A to keep both RH and LH printed parts in the cell at all times. This further reduces RH-to-LH changeover time at the pacemaker process, where very frequent changeovers' (leveling the mix) is a key lean objective.

Each container in the cell for example, a bin that holds 60 printed parts, or about one hour of current printing box assembly will have a withdrawal kanban with it. When a cell operator begins taking parts out of another bin, its withdrawal kanban is given to the material handler so that he/she knows to go to the printing supermarket and "withdraw" another bin of those parts.

Withdrawal kanban trigger the movement of parts. Production kanban trigger the production of parts. Company A can attach a production kanban to each bin of 1800 printed parts in the supermarket Every time the material handler removes a bin from the supermarket a kanban will be sent back to the printing machine. This instructs printing to produce 1800 parts, place them in a bin, and move it to a specified location in the printing supermarket.

Now printing no longer receives a schedule from production control. With mapping icons the flow can be seen as that show in Figure 5.6:

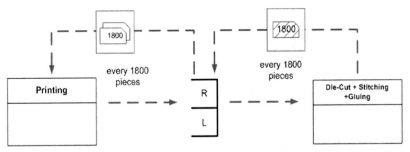

Figure 5.6   Flow of Products with Kanban System

Until changeover time on the printing press is greatly reduced, replenishing what is withdrawn from the printing supermarket on a bin-for-bin basis is clearly not practical.

Due to change over time, printing needs to produce batches larger than 1800 pieces between changeovers. With the initial goal of "every part every day," printing's target batch size for printing boxes would be approximately 16,555 RH and 11,036 LH pieces. Printing will keep 1.5 days of parts in its supermarket, one-half day extra to allow for replenishment delay and some printing problems.

So Company A will use a signal kanban to schedule printing. In this case the kanban for right- and left-drive parts is brought from the supermarket to the printing machine whenever the number of bins remaining in the supermarket drops to a trigger ("minimum") point. When a triangle kanban arrives at the printing machine' scheduling board, it initiates a changeover and production of a predetermined batch size of a specific part. Printing still does not receive a schedule from production control.

Drawn with icons, the flow now looks like that show in Figure 5.7:

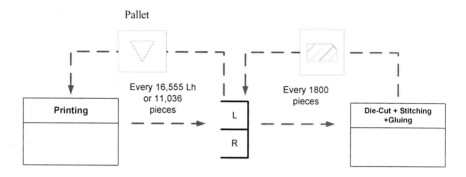

Figure 5.7   Final Flows of Products with Kanban System

The printed-parts supermarket, withdrawal and signal kanban, and kanban flows (dotted lines) are drawn on the future-state map.

5.6.2   Pallets of Paper

Future-state map must also show a third supermarket at the receiving dock, which holds pallets of paper. Even though Company A's paper supplier is not ready to receive kanban and produce according to them, Company A can still attach an internal withdrawal kanban to every pallet and send those kanban to its own production control department whenever another pallet is used. Production control can then order pallets based on actual usage, instead of based on MRP's best guess of what future usage will be.

Once production control has made the day's order for pallets, the corresponding kanban can be placed in kanban slots at the receiving dock. These indicate the day that pallets should arrive. If there are kanban still left in yesterday's receiving slot, then something is wrong at the supplier.

Currently the paper supplier is shipping pallets weekly. By lining up other customers along a "milk run" delivery, it may be possible to get the necessary amount of paper on a daily basis, even if the paper supplier does nothing to reduce its minimum batch size for pallets. Simply moving to daily delivery eliminates 80% of the inventory at Company A, while providing smooth, steady demand for the paper supplier.

5.6.3   The Progress So Far

It is now proposed a cell of the type many firms have implemented in the past few years, the introduction of pull to control printing production and pallet delivery, "every part every day" in printing, and instituting milk runs for delivery from the raw material supplier to Company A. By constructing a "before-and-after" in Table 5.1 for the current state and the future state so far, a large amount of waste can be observed and removed through these actions.

Table 5.1 : Company A Lead Time Improvement

	Pallet	Printing Parts	Die-Cut/Stitching/Gluing WIP	Finished Goods	Production Lead Time
Before	4 Days	3.9 Days	5.88Days	2.85Days	17 Days
Next	2.5Days	1.5 Days	-	2.85Days	7 Days

These are big steps forward. However, if the rest of the information flow at Company A is not fundamentally changed, it will be very difficult to operate a lean value stream. So it is necessary to go back to the customer and rethink the flow of information about customer desires as it is sent back to Company A and used there.

5.6.4   Second View of the Future-State Map

Figure 5.8 shows the second view of the Future-State Map showing stamping and raw material supermarkets

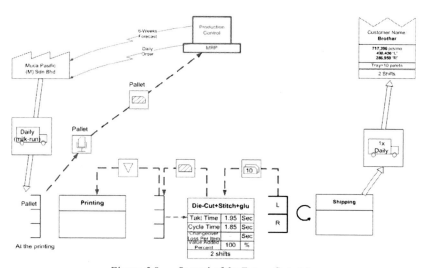

Figure 5.8   Second of the Future-State Map

Currently, the customer is sending by fax a 90-day forecast, revised once a month and frozen at 30 days. In addition, it is sending a daily release during the night by EDI (Electronic Data Interchange—in plain language, by phone line) to Company A's scheduling computer for the next day's shipping requirement. Finally, there are occasional revisions in shipping requirements on an emergency basis. These are sent by phone from the customer's material handling department to

Company A's shipping department during the day as the assembly plant discovers that needed parts are not on-hand for whatever reason.

What happens to the information sent from the customer once it reaches Company A? In the current case, the weekly schedule is fed over the weekend into the computerized MRP, which then sends instructions by Monday morning to each department printing, die-cutting, stitching and gluing about what to make the coming week. Then, as additional information is received each night and as each department reports back periodically to the MRP on what it actually did that day (because production does not go as scheduled), the daily production schedules are continually adjusted to bring what Company A is making into sync with what the customer wants.

If this sounds complicated it is because trying to run operations off of MRP systems doesn't work well. There is still a frequent need for humans to override the system to avoid shortages at various stages of production. The occasional call from the customer for emergency changes in orders requires human intervention as well and upsets the entire production schedule, requiring recalculation and retransmission to the processing areas.

5.7   Single point in the production chain

According to process steps downstream of the pacemaker process need to occur in a flow; in the Company A the scheduling point is clearly the die-cutting, stitching and gluing cell. We cannot schedule any further upstream (at the printing process) because we are planning to introduce a pull system between printing and die-cut, stitch and glue. This single scheduling point will regulate Company A's entire printing box value stream.

5.8   Level Production Mix at the Pacemaker Process

When the daily delivery is made to the assembly plant, 9 trays of right-drive brackets (16,555 pieces) and 6 trays of left-drive brackets (11,036 pieces) are typically staged and loaded onto the truck at one time. If not careful, the 15 production kanban removed from these trays before loading will be sent back to the die-cut, stitch and glue cell in a batch, as shown on the future-state map so far. If this happens the die-cut, stitch and glue cell will probably batch-produce these parts. That is, the cell will produce all 9 trays of right-drive brackets, and then change over to make the 6 trays of left-drive brackets, which would look like this:

1st Shift   2nd Shift

RRRRRRRRRRRRRRRRRRRRRRRRRLLLLLLLLLLLLLLLLLL

From the cell's perspective this seems to make sense because it minimizes the number of required set-up changes. However, from a value-stream perspective batching is the wrong way to go. Batch-producing brackets in assembly will increase the impact of problems, lengthen the lead time, and mean that the printed-parts supermarket has to be ready to meet sudden demand surges. "Being ready" means keeping more printed parts inventory in the supermarket, which again increases lead time, obscures printing's quality problems and, in general, causes all those wastes associated with overproduction.

Instead, if the die-cut, stitch and glue cell levels the mix of brackets it produces evenly over the shift, then the printing press (with shortened setup time) will have plenty of time to react to the cell's pulls for left-drive and right-drive parts. It will have time to replenish what was taken away without the need for so many inventories in the printing supermarket.

With leveling, this requires much more frequent changeovers. The cell's production mix of trays of brackets would look like this:

94

1st Shift   **2nd Shift**

LRRLRRLRRLRRLRRLRRLRRLRRLRRLRRLRRLRRLRRLRRLRRLRRLRRL

It can be ensured that kanban coming back to the die-cut, stitch and glue cell, which are the production instructions, come back in a sequence that levels the mix of products over the shift. At Company A there are two places where the batch of kanban can be intercepted and this leveling can take place. (it can be assumed that Company A has decided to use a load-leveling box to help maintain a level production mix, paced withdrawal, and genuine pull.)

Option A (Refer Figure 5.9).

Production control can place withdrawal ("move") kanban corresponding to the customer order in a load-leveling box near the shipping dock in a mixed, right-drive/left-drive sequence. A material handler then pulls these kanban out of the leveling box one-by-one at the pitch increment , and moves trays of brackets from the finished goods supermarket to the staging area one-by-one according to the withdrawal Kanban.

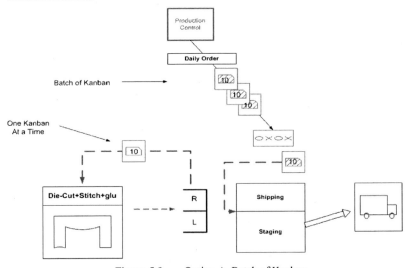

Figure 5.9   Option A, Batch of Kanban

As each tray is pulled from the supermarket, the production kanban on those trays are brought back to the cell in time increments and a right-drive/left-drive pattern that mirrors exactly the mix and pitch increment that production control had set up. (This leveling option is the one shown in Company A's completed future-state map.)

5.9   Load- Leveling Box

Company A will provide Takt image to the die-cut, stitch and glue cell, and frequently it will check production there. Returning all 15 kanban (two shifts work) to the cell at once would provide no takt image to the cell. Batching the volume of work instruction like this must be avoided. A natural increment of die-cutting, stitching and gluing work in Company A's case is the 1.95-second takt time × 15 pieces per tray = 30 minutes. This is the steering-bracket pitch, which corresponds to one kanban for one tray of 15 steering brackets.

Each column in Company A's load-leveling box represents a 30-minute pitch increment. The two rows are designated for right-drive and left-drive kanban. Every 30 minutes, a material handler brings the next kanban to the die-cut, stitch and glue cell and moves the just-finished tray of brackets to the finished-goods area. If a tray is not finished at the 30-minute pitch increment, then Company A knows there is a production problem that needs attention.

Company A load-leveling box for Printing boxes (Refer Figure 5.10)
Die-cut, stitch and glue cell gets kanban from left to right at pitch increment

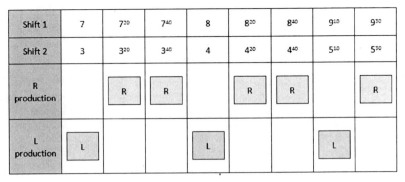

Shift 1	7	7^{20}	7^{40}	8	8^{20}	8^{40}	9^{10}	9^{30}
Shift 2	3	3^{20}	3^{40}	4	4^{20}	4^{40}	5^{10}	5^{30}
R production		R	R		R	R		R
L production	L			L			L	

Figure 5.10   Load-Leveling Boxes

5.10   Improvements process

Achieving the material and information flows it is envisioned for Company A. Printing requires the following process improvements:

- Reduction in changeover time and batch sizes at the printing press, to allow faster response to downstream usage. The goals are "every part every day" and then "every part every shift."

- Elimination of the long time (10 minutes) required to change between right-drive and left-drive fixtures in die-cutting, to make possible continuous flow and mixed production from die-cutting through assembly.

- Improvement in on-demand uptime of the second gluing machine, as it will now be tied to other processes in a continuous flow.

- Elimination of waste in the die-cut, stitch and glue cell, to reduce total work content down to 5.56 seconds or less (which allows use of 3 operators at the current demand level).

It is necessary to figure out how to use the existing printing technology-designed to produce printings in much higher volume than the customer for this product desires-in a less wasteful way. The secret here is to have the printing machine, which also prints parts for other product families in the plant make smaller batches of the two parts our value stream needs and make them more frequently. This will require additional reduction of the changeover time.

In fact, the methods for reducing setup times on a printing machine are well known and a reduction in time to less than 10 minutes can be achieved quickly. With that, it can be imagined the press making only about 16,555 right-drive printings and 11,036 left-drive printings; then producing parts for other value streams; then making more rights and lefts on the next shift.

This way the amount of inventory stored between the printing process and the die-cut, stitch and glue cell would be reduced.

However, be sure to kick off these improvement projects by creating a "pull" for the improvements. That is, instead of "pushing" a team to reduce setup time on the printing press, begin instead by stating that in 30 days the batch sizes on the printing machine will be reduced to 16,555 & 11,036 pieces. This creates a sense of urgency about making the process improvement. Likewise, don't simply send a team to eliminate the set-up changeover time and wait for them to be finished. Begin by stating that in 14 days the die-cutting, stitching and gluing steps will be placed into a continuous flow orientation.

98

Draw the complete future-state value-stream map,can be drawn for Company
A with information flows, material flows, and kaizen needs specified as show in
Figure 5.11.

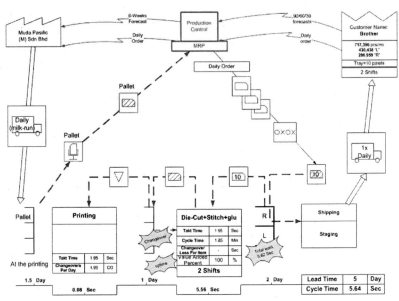

Figure 5.11   Final Future-State Mapping

5.11   Summary

Comparing the summary statistics for Company A's current state and its
future state, the results are quite striking. In particular, due to leveling production in
its die-cut, stitch and glue cell and developing the ability to print every part every
shift, Company A can further reduce the amount of pallets and printed parts held in
supermarkets. Of course, this puts great pressure on maintaining equipment
reliability and predictability of production to takt.

With the shortened production lead time through its shop floor, the pacemaker process operating consistently to takt time, and fast response to problems, Company A can comfortably reduce the amount of finished goods it holds to two days. (If Company A's customer were to level its schedule, this finished-goods inventory could be reduced even further.) Refer Figure 5.2

Table 5.2 : Company A Final Lead Time Improvement

	Pallet	Printing Parts	Die-Cut/Stitching/Gluing WIP	Finished Goods	Production Lead Time
Before	4 Days	3.9 Days	5.88Days	2.85 Days	17 Days
Continuous flow & pull	2.5 Days	1.5 Days	-	2.85 Days	7 Days
With Leveling	1.5 Day	1 Day	-	2 Days	5 Days

5.12 Information Modeling

5.12.1 Entity Relationship Diagram (IDEF1X)

IDEF1X is a data modeling language for the developing which relates data models. It is used to produce graphical information model which represents the structure and relate of information within an environment or system.

Refer is to Figure 5.12, the production plan departments only connect with the last station which is the product delivery. After the product delivery receive the kanban cards from production plan department and proceed the data, then data and production information goes to other work station respectively until the start of process

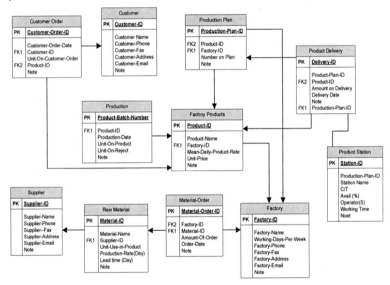

Figure 5.12   Entity Relationship Diagram (IDEF1X)

5.12.2 Data Relationship

With refence to Figure 5.13, the data relationship are shown

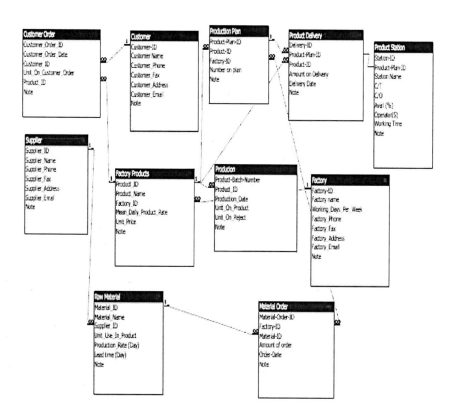

Figure 5.13 Data Relationship

5.13 Concluding

Value Stream improvement is work primarily a management responsibility. Management has to understand that its role is to see the overall flow, develop a vision of an improved lean flow for the future, and lead its implementation. We can ask the front lines to work on eliminating waste, but only management has the perspective to see the total flow as it cuts across departmental and functional boundaries.

CHAPTER 6

CONCLUSION

6.1   Introduction

This chapter provides a summary of the entire study. Some interesting findings from the study will then be addressed. Recommendations for future studies will be given as well. Lastly, the chapter ends with a final conclusion on the study.

6.2   Impact of Lean implementation

From the results and discussion it can be concluded that lean concept borrowed from Lean Manufacturing or Toyota Production system can be successfully implemented in service sectors.

6.3 Challenges in Implementing Lean

Lean implementation in many companies has failed despite effective use of some or all lean tools. Lean is often considered as a set of tools instead of a culture due to which many lean improvement programs either fail or do not reach the desired results. Some of the common reasons for lean failures are listed below:

1) Lack of management support2) Resistance to change (lack of buy in) from supervision and workforce3) Poor metrics 4) Not enough training 5) Little or no impact on profitability 6) Ineffective communications 7) Not able to sustain initial efforts 8) Not expanding improvements from the initial efforts to other departments 9) Improvements in one area seemed to have negative impact in other sections.

Lean improvement programs have to be tied up at a strategic level in an organization rather than a mere tactical level. Regardless of what lean tools or approaches are used each must be modified or customized using these 4 principles which return back to strategic planning of the business. The overall strategy must be based on these 4 key principles which are listed below

1) Not involving stakeholders in the planning process (primary cause for resistance to change) 2) Not considering the entire organization as a system 3) Not treating each company and each lean initiative as unique 4) Not understanding the real purpose for the initiative.

6.4 Project Summary

The purpose of this study was to improve printer boxes department operations to be leaner through the use of value stream mapping.

This chapter has analyzed whether lean production philosophy can be applied to the supply chain. It has shown that it is possible as long as a number of conditions exist. Specifically, the entire supply chain for each construction project needs to be managed in such a way that alliances are formed within it, with knowledge and experience being shared and with everyone working towards a single common goal: customer satisfaction. In this way it will be possible to shorten delivery times on the project, reduce costs and increase quality, thereby increasing the satisfaction of both internal and external customers and improving the effectiveness and efficiency of the process as a whole.

A current state VSM was developed and analyzed for potential areas of improvement. The current state map laid out the assembly processes, together with the work in queue level, value-added time, non-value-added time, and other information of the line production. From the current state VSM, it was clearly shown that the production line has a long lead time and low value-added ratio. The causes of the long lead time and low value-added ratio are push production system and unbalanced processing times between the operators.

6.5   Findings

Throughout the project, value stream mapping has proven to be an outstanding tool to analyze production systems. Current state VSM helps to identify areas of potential improvement; while future state VSM helps the system users to visualize the state of the production upon successfully implementing the improvements.

A weakness of value stream mapping was addressed in this project. VSM does not show the dynamic behavioral aspects of production systems and most of the time the future state VSM is developed based on experience without concrete evidence. Therefore, sometimes, future state VSM is viewed as unrealistic to achieve; and the entire improvement project may end up in failure.

One interesting finding in the project is that indeed balancing the lines could reduce the number of operators needed and increase the throughput of the system; but maintaining a push production system could not promise significant improvements in terms of lead time, WIP level, work-in-queue level, product waiting time, and value added ratio.

Therefore, it can be said that the primary cause of long lead time, high WIP level, high work-in-queue level, high time in queue of products, is push production system, instead of unbalanced lines.

However, the necessities and benefits of line balancing should not be neglected or underestimated. Pull production system with line balancing, it shows that implementing both line balancing and pull production system in the assembly department could achieve the greatest improvements for the system. This is reasonable and complying with the lean concept which suggests that clearing out the obstructions in the flow of product would build a solid foundation upon which to create a kanban-based pull production system. In other word, balancing the lines helps to clear out the obstructions in the product flow and this will form a strong foundation for the pull production system to success.

6.6   Future Research

The study was carried out to establish improvements that would help the printer Boxes production line to become leaner. In this study, value stream mapping was implemented in Company A and these are found to be an excellent combination of tools to evaluate the current situation of a system, identify areas for improvements and evaluating different improvement alternatives.

Therefore, it is recommend that the future works of mapping other products lines in Company A to create a better model of the entire plant floor. Further studies should be conducted in addition to this study in order to develop a larger model of the company. Furthermore, the map can be extended further to include other aspects of the company, such as SCM and quality management. Further it is important to validate the future state map using simulation software such as Witness and try to develop a database to simplify the flow of information and using online work order for customer's demand to avoid working time constrains

6.7   Conclusion

This study was carried out on Printer Boxes line production. The study was conducted successfully. Current state VSM was developed to identify the wastes and possible improvement areas in the system. Improvement alternatives were developed and compared using information model. And lastly, a future state VSM was developed. The combination of value stream mapping and information system was proven to be an excellent tool in identifying wastes and areas for improvements, documenting details of the system, propose improvement alternatives, and evaluating the improvement alternatives. It is believed that if the improvements suggested in the study are successfully implemented by Company A, the expected outcomes will be achieved and the assembly department will become leaner

REFERENCES

Allen, B. R., and Boynton, A. C. (1991) Information Architecture: In Search of Efficient Flexibility, MIS Quarterly, 435-445.

Chandra Charu &GrabisJānis (2007). Supply Chain Configuration. United States of America: Springer US

Chopra, S., & Meindl, P. (2001). Supply Chain Management: Strategy, planning and operation.: Prentice Hall.

Dagenais, T., and Gautschi, D. (2002) Net Markets: Driving Success in the B2B Networked Economy, McGraw-Hill Ryerson Ltd., Toronto, ON.

DNREC (2005). DNREC/DEDO Statement Regarding Value Stream Mapping Workshops : VSM in Delawe. Retrieved February 19, 2009, from http://www.dnrec.state.de.us/DNREC2000/VSM/Index.htm

Gidley, S. (2004). The Tools of Lean - Value Stream Mapping. Mahoney: Institute of Business Excellence.

Henderson, J. C., Venkatraman, N., and Oldach, S. (1996) Aligning Business and IT Strategies. In: J. N. Luftman (ed.), Competing in the Information Age: Strategic Alignment in Practice, Oxford University Press, New York, NY, 21-41.

Kearns, G. S., and Lederer, A. L. (2001) Strategic IT Alignment: A Model for Competitive Advantage, Proceedings of the Twenty-Second International Conference on Information Systems, New Orleans, LA, December 16-19

Levinson, W., & Rerick, R. (2002). Lean Enterprise: A Synergistic Approach to Minimizing Waste. Milwaukee, WI: ASQ Quality Press.

Lee, H. G., Clark, T., and Tam, K. T. (1999) Research Report. Can EDI Benefit Adopters?, Information Systems Research, 10(2), 186-195.

Lucas, H. C. (1981) Implementation: The Key to Successful Information Systems, Columbia University Press, New York, NY.

Melton, P.M., 2004, To lean or not to lean? (that is the question), The Chemical Engineer, September 2004 (759): 34–37.

Ohno T. Toyota Production System: Beyond Large-Scale Production. Portland, OR: Productivity Press; 1988.

Reddy, R., and Reddy, S. (2001) Supply Chains to Virtual Integration, McGraw-Hill, New York, NY.

Santos, J., Wysk, A.R., Torres, M.J. (2006). Improving production with lean thinking. New Jersey: John Wiley & Sons, Inc.

Sayer, N.J., and Williams, B. (2007). Lean for Dummies. Indiana: Wiley Publishing, Inc.

Ulrich D. and Lake, D., 1990, Organisational Capability: Competing from the inside/out. New York: Wiley

Womack, P.J., Jones, T.D., Roos, D. (1990). The Machine that Changed the World: The Story of Lean Production. New York: Harper Perennial.

Womack, J.P. and Jones, D.T. Lean Thinking: Banish Waste and Create Wealth in Your Corporation, Simon & Schuster, New York, USA (1996).